Climate Change Refugees

Julia Wall

NELSON
CENGAGE Learning

Australia • Brazil • Japan • Korea • Mexico • Singapore • Spain • United Kingdom • United States

Climate Change Refugees

Text: Julia Wall
Editor: Rebecca Crisp
Design: James Lowe
Series design: James Lowe
Photo researcher: Lisa Piemonte
Production controllers: Lisa Porter and Renee Cusmano

Acknowledgements
The author and publisher would like to acknowledge permission to reproduce material from the following sources:
Corbis Australia: pp. 1, 5 (bottom), cover; David Liam Kyle/NBAE via Getty Images: p. 23 (bottom); Getty Images: pp. 3, 4 (bottom), 5 (top), 6, 14, 15, 22 (both), 23 (top); Getty Images/AFP/Pascal Guyot: p. 4 (top); Grant Reed © Cengage Learning Australia: pp. 8, 10, 17; iStockphoto: p. 12 (bottom); Jeremy Sutton-Hibbert/WpN/Picture Media: pp. 9, 11; Mark Lynas/Still Pictures: p. 16; Peter Bennetts/Lonely Planet Images: p. 18; Photolibrary: pp. 7, 19, 20, 21; Shutterstock/Ben Heys: pp. 13, back cover.

Every effort has been made to trace and acknowledge copyright. However, if any infringement has occurred, the publishers tender their apologies and invite the copyright holders to contact them.

Fast Forward Independent Texts
Level 15

ISBN 978 0 17 017950 8
ISBN 978 0 17 017897 6 (set)

Cengage Learning Australia
Level 7, 80 Dorcas Street
South Melbourne, Victoria Australia 3205
Phone: 1300 790 853

Cengage Learning New Zealand
Unit 4B Rosedale Office Park
331 Rosedale Road, Albany, North Shore NZ 0632
Phone: 0800 449 725

For learning solutions, visit **cengage.com.au**

Printed in Australia by Ligare Pty Ltd
4 5 6 7 8 9 10 20 19 18 17 16

Climate Change Refugees

Julia Wall

Contents

Chapter 1 Refugees 4
Chapter 2 Climate Change Refugees 6
Chapter 3 The Disappearing Islands 8
Chapter 4 Climate Change 12
Chapter 5 Refugee Hot Spots 15
Chapter 6 Starting Again 22
Glossary and Index 24

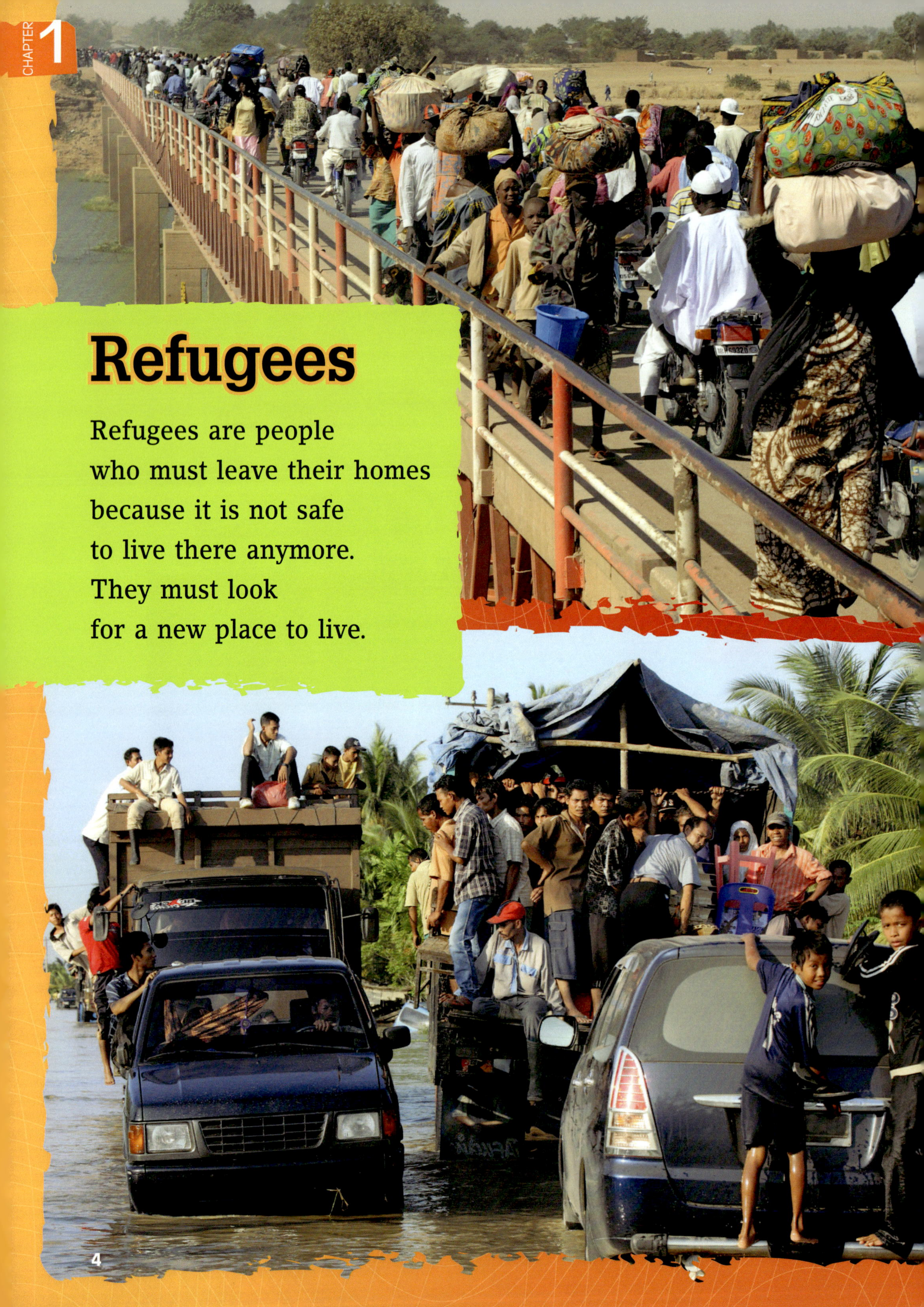

CHAPTER 1

Refugees

Refugees are people who must leave their homes because it is not safe to live there anymore. They must look for a new place to live.

There are different reasons why people become refugees. Some people leave their homes because there is a war.

Some people leave because other people in their country want to harm them.

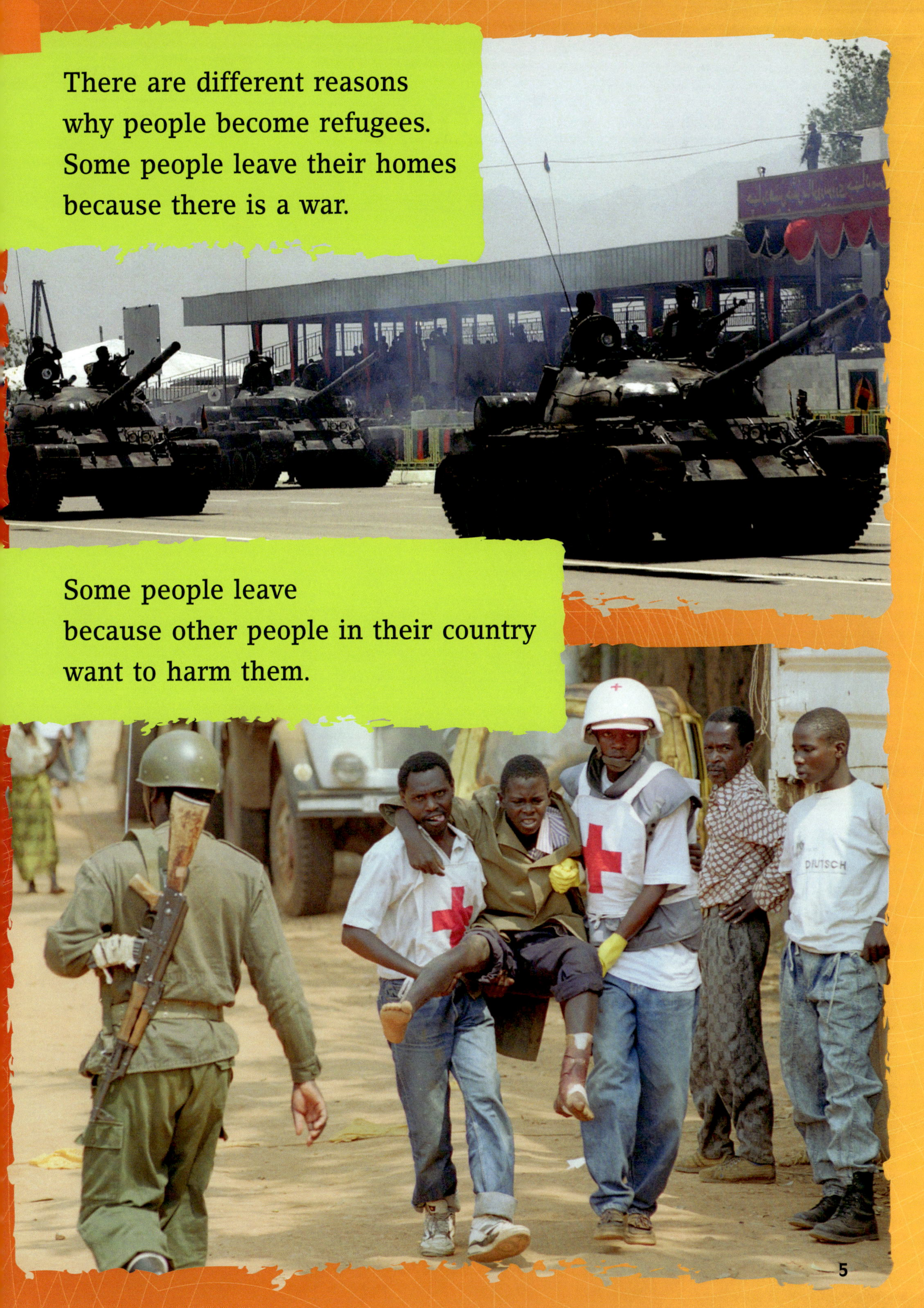

CHAPTER 2

Climate Change Refugees

Climate change refugees are people who leave their homes because there has been a big change in the climate.

The change may have happened suddenly, such as a **tsunami** or a big storm.

Around 200000 people became refugees when Hurricane Katrina hit the USA in 2005.

The change in the weather may have happened over some time.
There might have been too little rain for many years.

The Disappearing Islands

The first climate change refugees used to live on the Carteret Islands.

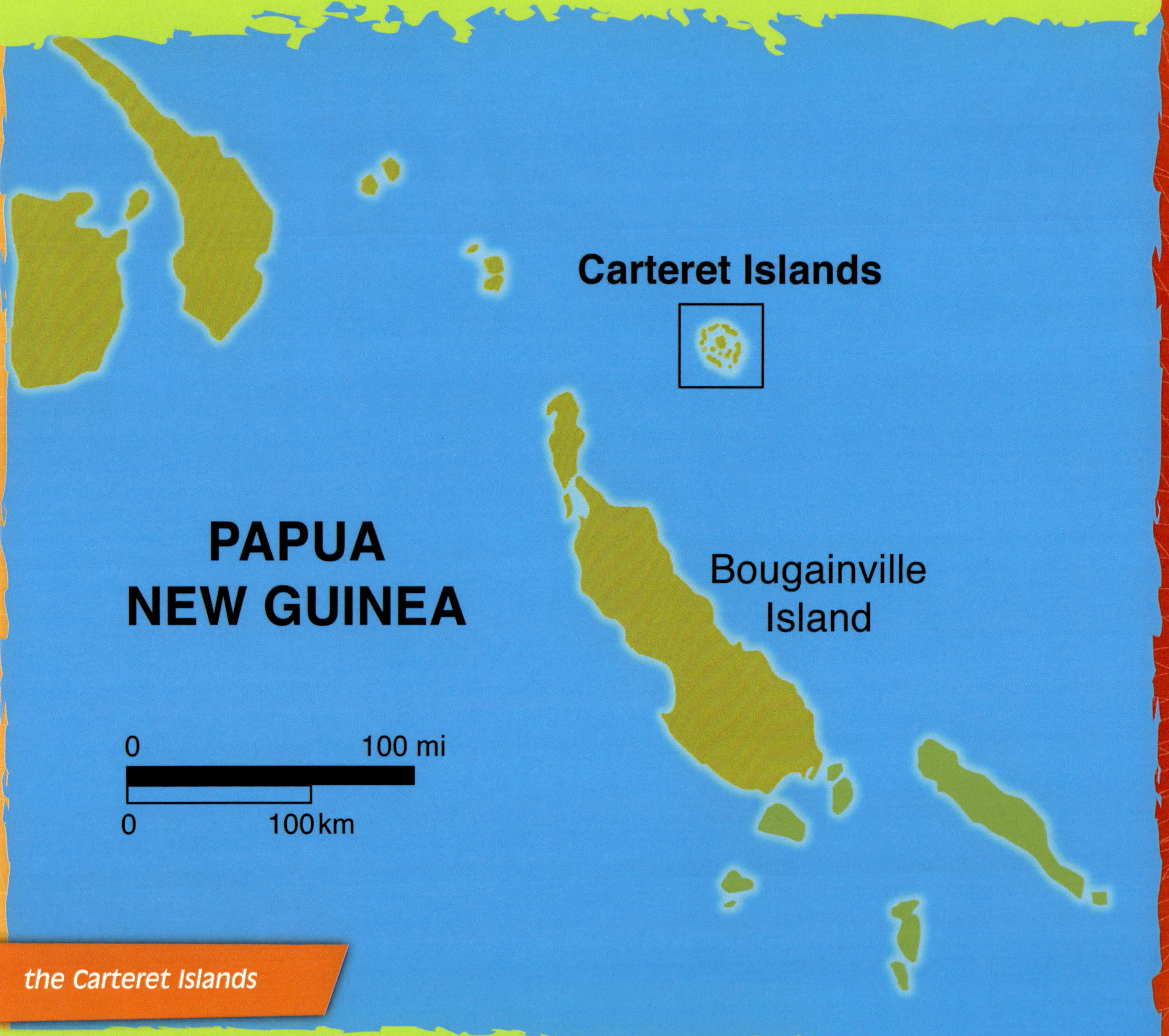

the Carteret Islands

Over many years,
storms and high **tides** harmed homes,
crops and drinking water.

The people who lived there tried
to block the rising sea with trees.
They built a wall too,
but the sea was too high.

a beach on the Carteret Islands

The people started to run out of food.
Their home was disappearing
under the sea,
so the people started to leave.

People from the Carteret Islands have had to move to Papua New Guinea.

Scientists think that the Carteret Islands will all be underwater within a few decades.

Climate Change

Today, there are more climate change refugees than refugees who leave their homes because of war or other reasons.

What Causes Climate Change?

Some scientists say that pollution causes climate change. Pollution causes **global warming**, which causes changes in the world's climate.

Pollution makes gases that trap heat from the Sun in Earth's atmosphere.

Global warming is changing the climate. The world is getting hotter. This will cause more disasters in the **environment**, so there will be more and more climate change refugees in the future.

Australia is currently experiencing a drought.

As the world gets hotter,
crops will die.
People will have less food to eat.
They will have to leave their homes
to find new places to grow food.

This is already happening
in some places.

These people are lining up for food because their crops died in a drought.

5

Refugee Hot Spots

In the future,
climate change refugees will come
from all over the world.
There are more disasters
and dry places today
than ever before.

*This house in China was damaged by a **landslide**.*

The sea is getting higher all over the world and land is disappearing under the water.

Every year, the sea reaches further inland.

The sea water gets into the fresh water, which harms drinking water and water for crops.

Tuvalu

In Tuvalu, the sea is rising fast
and **poisoning** the land
and the trees.

Climate change is heating the sea water
and harming the coral
where fish live.
So there are not as many fish
for people to eat.

The islands of Tuvalu are north-east of Australia.

Many people from Tuvalu have become climate change refugees.
They have had to move to New Zealand.

former citizens of Tuvalu

In the Mountains

In years to come,
some people who live near mountains
may also become climate change refugees.

As Earth gets hotter,
snow at the top of these mountains will **melt**.
A huge amount of water will run down
to where the people live.

Melting Snow

As the snow on this mountain melts, the water flows into the lake. When the lake gets too full, the water will run down to the villages below.

These people will have to move to lower ground to find safer places to live.
Their houses will be washed away by the melted snow.

CHAPTER 6

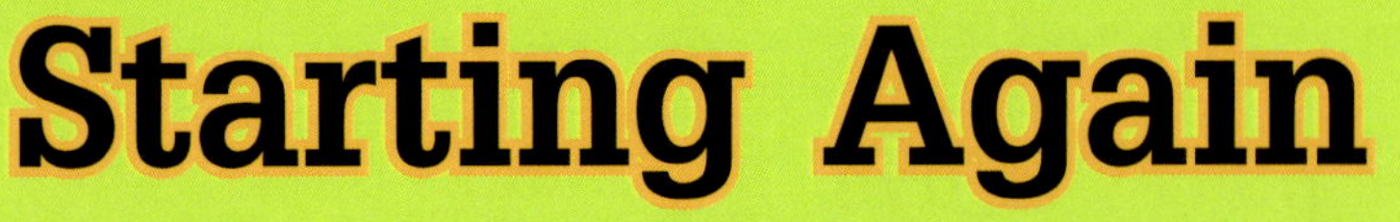

Starting Again

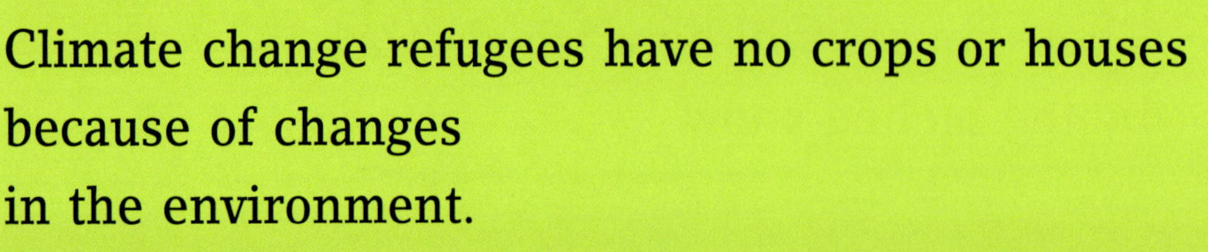

Climate change refugees have no crops or houses because of changes in the environment.

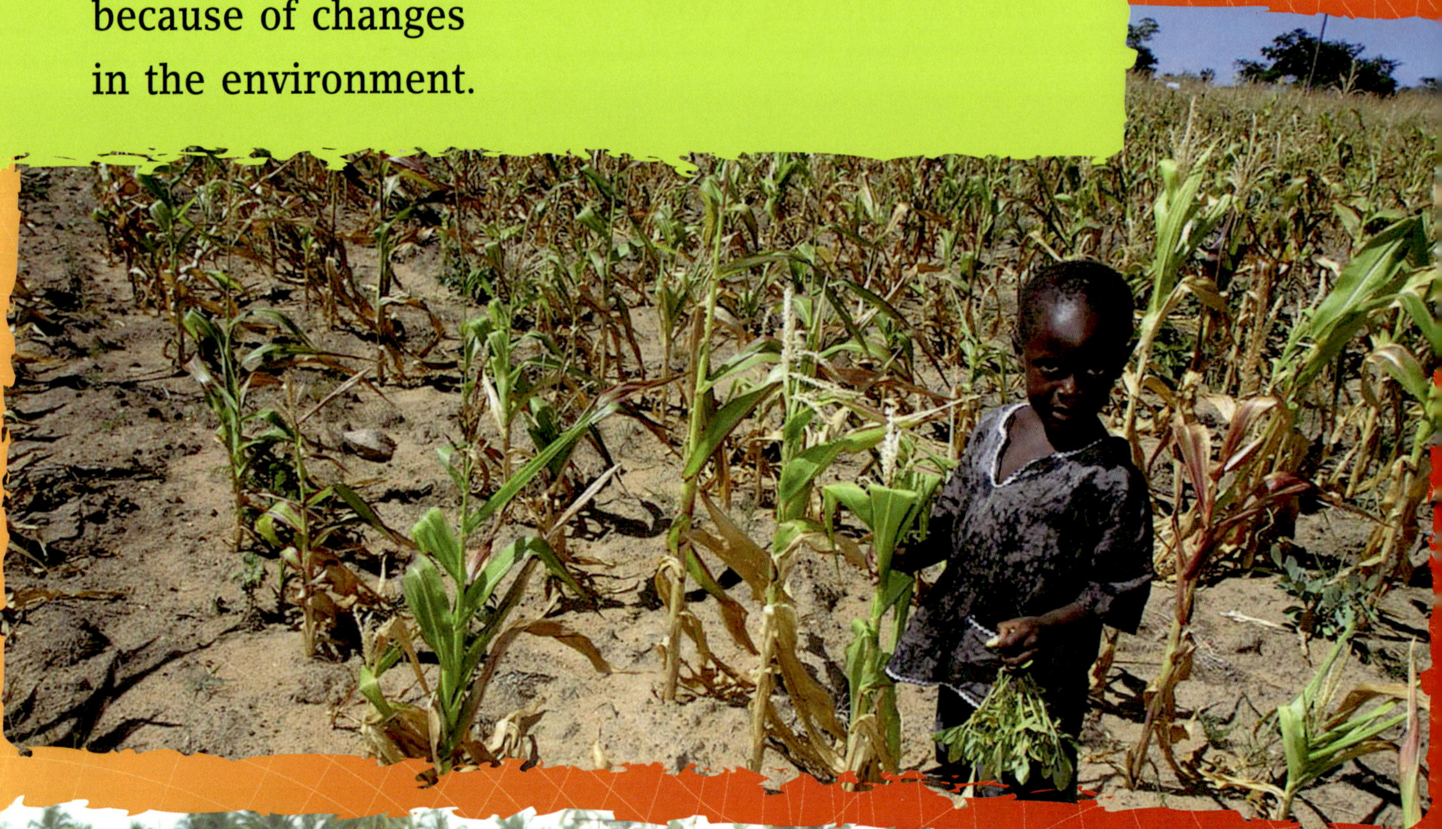

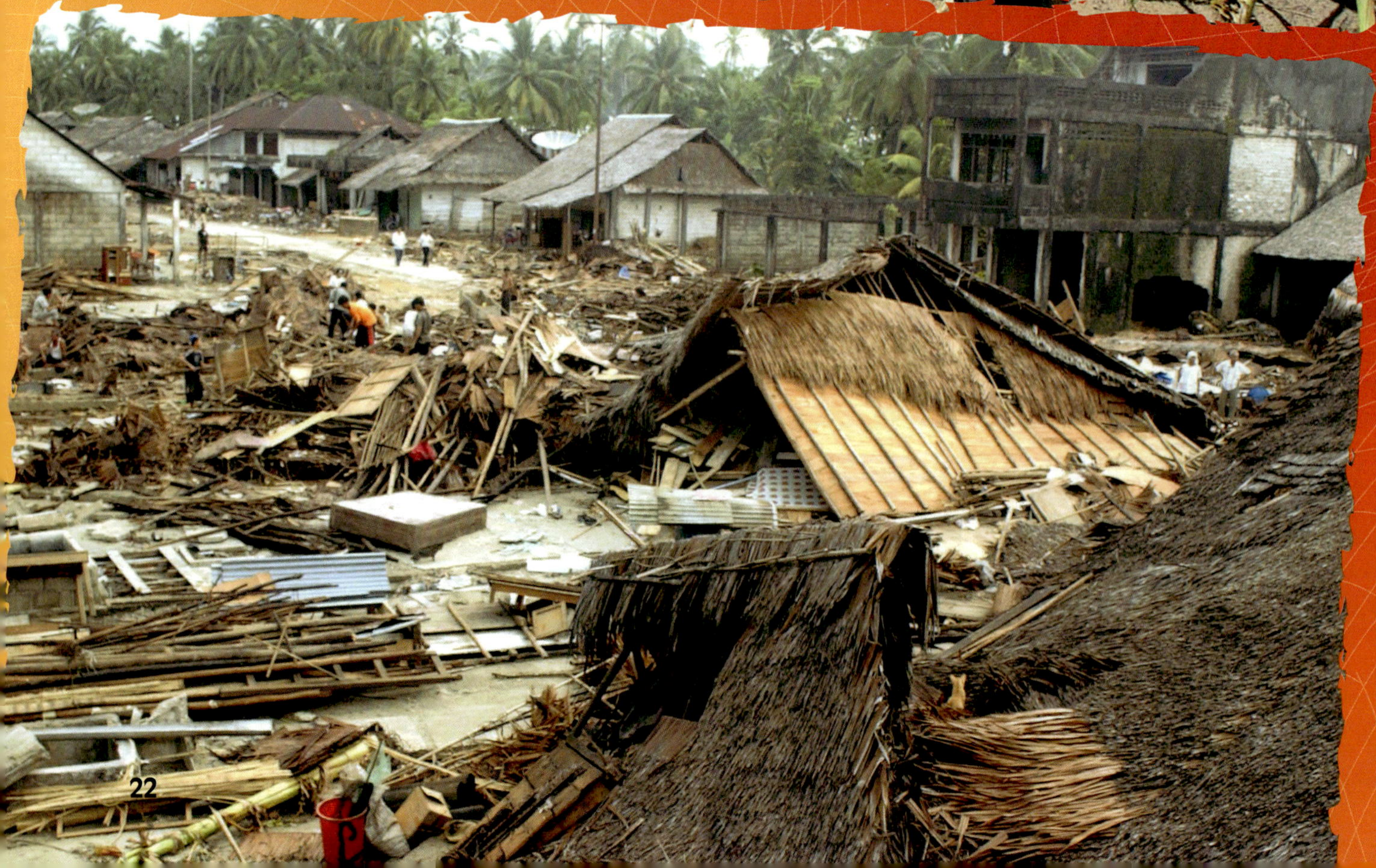

They will need homes and food, and help to start again in new places.

Everyone should try to look after the environment so there will not be as many climate change refugees in the future.

Glossary

climate	the weather conditions in an area over time
environment	the surrounding area
global warming	the gradual rise in Earth's temperature
landslide	a huge flow of mud, earth and rocks down a hillside, sometimes caused by heavy rain
melt	become liquid
poisoning	using something to kill or harm
tides	the regular rising and falling of the level of the ocean, caused by the rising and setting of the Moon
tsunami	a huge wave that wipes out coastal areas

Index

Carteret Islands 8–11
crops 8, 14, 16, 22
environment 13, 22–23
food 10, 14, 23
global warming 13
mountains 19–20
New Zealand 18
tsunami 6
Tuvalu 17–18
war 5, 12